LEARN ABOUT

POLYNOMIALS

IN

8 MINUTES

by

Arlissa Pinkelton

ISBN-13:
978-1512189575

Mathematical content edited by:
Harvey "Boe" Marshall

Cover Layout and Design by:
Tanika Ault
Legacy Designs by Tanika Ault

Dedication

To my Mommy, Willow Dean, thanks for always encouraging me to learn more, do more, and be more.

About The Author

Arlissa Pinkelton is a Mechanical Engineer turned Mathematics professor. She lives with her family in central Florida where she teaches Algebra. Arlissa has loved math since childhood and seeks to inspire other students with that same passion. At the time of this printing, she'd taught math for 9 years. Having taught hundreds of students, she loves providing tools and resources to improve math success.

Foreword

Learn Polynomials In 8 Minutes is the third book in a series of *"8 Minute"* books on topics in Algebra. These titles are designed to make learning Algebra easier and quicker to learn.

Other Titles

Learn To Factor in 8 Minutes

Learn Exponents in 8 Minutes

Upcoming Titles

Learn Radicals in 8 Minutes

Solving & Graphing Linear Equations in 8 Minutes

Learn All You Need To Know About Real Numbers in 8 Minutes

Learn Rational Expressions in 8 Minutes

Learn About Inequalities in 8 Minutes

Table of Contents

LEARN POLYNOMIALS IN 8 MINUTES

Introduction

A **Polynomial** in one variable is defined as a single term or a finite sum of terms of the form ax^n, where a is the real number, x is the variable, and n is the non-negative exponent. It is an expression or equation in the form of $P(x) = a_n x^n + a_{n-1} x^{n-1} + \ldots + a_1 x + a_0$

Polynomials can be added, subtracted, multiplied, and divided. These can be done using basic operations and utilizing the skills you learned while working with exponents. If you're not comfortable working with exponents, you should first get and read my book *Learn Exponents in 8 Minutes.*

Polynomials are used in calculus and numerical analysis to approximate other functions; they appear in settings ranging from basic chemistry and physics to economics and social science. In advanced mathematics, polynomials are used to construct polynomial rings and algebraic varieties, central concepts in algebra and algebraic geometry.

8 Sub-topics in Polynomials

1. **Adding Polynomials**

2. **Subtracting Polynomials**

3. **Multiplying by Monomials**

4. **Multiplying by Polynomials**

5. **Special Products (Difference of Squares, Perfect Square Trinomials, and Cubics)**

6. **Dividing by Monomials**

7. **Dividing by Polynomials including Long Division**

8. **Applications using Polynomials**

Terminology

For each term of a polynomial, ax^n, a is called the **coefficient**, x is the **variable**, and n is the **exponent** and is also called the **degree of the term**.

For example,

$-12x^7$	Coefficient: -12	Degree: 7
x^3	Coefficient: 1 *(invisible coefficient)*	Degree: 3
$10w$	Coefficient: 10	Degree: 1 *(invisible exponent)*

Polynomials are a series of terms put together by mathematical operations. A polynomial with exactly one term is called a **monomial**. A two-term polynomial is a **binomial**. A polynomial with three terms is known as a **trinomial**. Polynomials can be one term, two terms, three terms, or longer. When there are more than three terms, there's no specific name for them.

Parts of a Polynomial

Polynomials are generally written with the degrees in descending order. The term with the highest degree is called the **leading term** with its coefficient being called the **leading coefficient**. The highest degree of all of the terms in the polynomial is called the **degree of the polynomial**.

Polynomials may have more than one variable in a term. In this case, the **degree of the term** is the sum of the exponents in that term. As an example, $3x^2y^4$ has a degree of 6 based on the adding of the exponents 2 and 4.

Examples:

	Type	Leading coefficient	Degree of polynomial
$-3x^4$	Monomial	-3	4
25	Monomial	25	0
$-4x^4y^3z^2$	Monomial	-4	9
$-6y^5 + 4y^3$	Binomial	-6	5
$8p^6-3p^3+4p$	Trinomial	8	6

Before we go any deeper into polynomials, we must refresh our minds about like terms and unlike terms. Terms that are *Like* have the exact same variable(s) and exact same exponent(s). If they don't have either of these exactly the same, then they are *Unlike* terms.

Like terms: $3x^2$, $-7x^2$ or $-5yz^3$, yz^3

Unlike terms:	$9z^3$, $12z^6$	Different exponents
	$-2m^4$, $3n^4$	Different variables
	$4y$, 7	Different variables

Adding Polynomials

When adding polynomials, you must first identify the like terms. Once you've done that, add the coefficients of these terms and leave the variables as they are. *Note: You cannot add terms that are Unlike.*

Example 1: $3x^3 + 7x^3 - 2x^3$

$$= (3 + 7 - 2)x^3 \text{ (notice keeping the variable as is)}$$

$$= (8)x^3$$

$$= 8x^3$$

Example 2: $4x^2 - 6y^4 - 3x^2 + 2y^4$

$$= 4x^2 - 3x^2 - 6y^4 + 2y^4 \text{ (Rearrange the}$$
$$\text{variables so the like terms are together)}$$

$$= (4 - 3)x^2 + (-6 + 2)y^4$$

$$= (1)x^2 + (-4)y^4$$

$$= x^2 - 4y^4 \text{ (*Notice how the unlike terms are left}$$
$$\text{as a polynomial in the answer.)}$$

Here are 8 practice problems for you to complete:

Evaluate each expression.

1   $3x^2y + 5x^2y$

2   $13a^2b^3 + 2a^2b^3$

3   $-5ab^3 + 17ab^3$

4   $(-3c^3 + 5c^2 - 7c) + (11c^3 + 6c^2 + 3)$

5   $(4w^2 - 2x) + (3w^2 - 4x^2 + 6x)$

6   $(5x^2 - 4xy + y^2) + (-3x^2 - 5y^2)$

7   $(\frac{5}{9}x - \frac{1}{10}y) + (-\frac{4}{9}x + \frac{3}{10}y)$

8   $(7.9t^3 + 2.6t - 1.1) + (-3.4t^2 + 3.4t - 3.1)$

Answers with explanations:

1   Answer: $8x^2y$

   $(3 + 5) x^2y = (8) x^2y = 8x^2y$

2   Answer: $15a^2b^3$

   $(13 + 2) a^2b^3 = (15) a^2b^3 = 15a^2b^3$

3   Answer: $12ab^3$

   $(-5 + 17) ab^3 = (12) ab^3 = 12ab^3$

4   Answer: $8c^3 + 11c^2 - 7c + 3$

   Put the like terms together:

   $(-3c^3 + 11c^3) + (5c^2 + 6c^2) + (-7c) + (3)$

   Add the coefficients and keep the variables:

   $= (-3 + 11)c^3 + (5 + 6)c^2 + (-7c) + (3)$

$$= (8)c^3 + (11)c^2 + (-7c) + (3)$$
$$= 8c^3 + 11c^2 - 7c + 3$$

5   Answer: $7w^2 - 4x^2 + 4x$

$$(4w^2 + 3w^2) + (-4x^2) + (-2x + 6x)$$
$$= (4 + 3)\, w^2 + (-4x^2) + (-2 + 6)\, x$$
$$= (7)\, w^2 + (-4)\, x^2 + (4)\, x$$
$$= 7w^2 - 4x^2 + 4x$$

6   Answer: $2x^2 - 4xy - 4y^2$

$$(5x^2 - 3x^2) + (-4xy) + (y^2 - 5y^2)$$
$$= (5 - 3)\, x^2 + (-4xy) + (1 - 5)\, y^2$$
$$= (2)\, x^2 + (-4)\, xy + (-4)\, y^2$$
$$= 2x^2 - 4xy - 4y^2$$

7   Answer: $\dfrac{1}{9}x + \dfrac{2}{5}y$

$$(\tfrac{5}{9}x + \tfrac{1}{10}y) + (-\tfrac{4}{9}x + \tfrac{3}{10}y)$$
$$= (\tfrac{5}{9} - \tfrac{4}{9})\, x + (\tfrac{1}{10} + \tfrac{3}{10})\, y$$
$$= \tfrac{1}{9}x + \tfrac{4}{10}y \quad \text{(Reduce to lowest terms)}$$

12

$$= \frac{1}{9}x + \frac{2}{5}y$$

8 Answers: $7.9t^3 - 3.4t^2 + 6t - 4.2$

$\qquad (7.9)t^3 + (-3.4)t^2 + (2.6 + 3.4)t + (-1.1 - 3.1)$

$\qquad = (7.9)t^3 + (-3.4)t^2 + (6.0)t + (-4.2)$

$\qquad = 7.9t^3 - 3.4t^2 + 6t - 4.2$

Subtracting Polynomials

Subtracting polynomials is much like adding them. In the case of subtracting, you must work with the opposites of each term. For example, subtracting 5x is the same as adding the opposite of 5x, which is –5x.

So, $A - B = A + (-B)$

Example 1: $3x - 5x = 3x + (-5x) = [3 + (-5)]x = -2x$

For some people, the concept of adding the negative is confusing. It may be best for you if you see it as (3 – 5)x. When you subtract 5x from 3x, the result is –2x.

When subtracting a quantity (or terms within parentheses), you must first distribute the – to each term in the quantity it precedes.

Example 2: $(2x - 4) - (5x + 10)$

$\qquad$ *Multiply – times 5x and 10*

13

$$= 2x - 4 - 5x - 10$$

Place like terms together

$$= (2x - 5x) + (-4 - 10)$$

$$= (2 - 5)x + (-4 - 10)$$

$$= (-3)x + (-14)$$

$$= -3x - 14$$

Here are 8 practice problems for you to try.

Evaluate each expression.

1    $4a^3b^2 - 12a^3b^2$

2    $-23c^5 - 12c^5$

3    $(4x + 3v) - (-3x + v)$

4    $(3h - 15) - (8h + 13)$

5    $(-4p^4 + 5p^2 - 3) - (11p^2 + 4p - 6)$

6    $(10r - 6s + 2t) - (12r - 3s - t)$

7    $(\frac{7}{8}x + \frac{2}{3}y - \frac{3}{10}) - (\frac{1}{8}x + \frac{1}{3}y)$

8    $(1.3c^3 + 4.8) - (4.3c^2 - 2c - 2.2)$

Answers:

1 Answer: $-8a^3b^2$

$$(4 - 12)a^3b^2 = -8a^3b^2$$

2   Answer: $-35c^5$

$$(-23 - 12)c^5 = -35c^5$$

3   Answer: $7x + 2v$

$$4x + 3v + 3x - v$$

Put the like terms together:

$$= (4x + 3x) + (3v - v)$$

$$= (4 + 3)x + (3 - 1)v$$

$$= 7x + 2v$$

4   Answer: $-5h - 28$

$$3h - 15 - 8h - 13$$

$$= (3h - 8h) + (-15 - 13)$$

$$= (3 - 8)h + (-15 - 13)$$

$$= (-5)h + (-28)$$

$$= -5h - 28$$

5   Answer: $-4p^4 - 6p^2 - 4p + 3$

$$(-4p^4 + 5p^2 - 3) - (11p^2 + 4p - 6)$$

Distribute the – to 11p², 4p, and –6:

$$= -4p^4 + 5p^2 - 3 - 11p^2 - 4p + 6$$

Put the like terms together:

$$= (-4p^4) + (5p^2 - 11p^2) + (-4p) + (-3 + 6)$$

$$= (-4p^4) + (5 - 11)p^2 + (-4)p + (-3 + 6)$$

$$= -4p^4 - 6p^2 - 4p + 3$$

6   Answer: $-2r - 3s + 3t$

$$10r - 6s + 2t - 12r + 3s + t$$

$$= (10r - 12r) + (-6s + 3s) + (2t + t)$$

$$= (-2r) + (-3s) + (3t)$$

$$= -2r - 3s + 3t$$

7   Answer: $\dfrac{3}{4}x + \dfrac{1}{3}y - \dfrac{3}{10}$

$$(\dfrac{7}{8}x + \dfrac{2}{3}y - \dfrac{3}{10}) - (\dfrac{1}{8}x + \dfrac{1}{3}y)$$

$$= \dfrac{7}{8}x + \dfrac{2}{3}y - \dfrac{3}{10} - \dfrac{1}{8}x - \dfrac{1}{3}y$$

$$= (\dfrac{7}{8} - \dfrac{1}{8})x + (\dfrac{2}{3} - \dfrac{1}{3})y - \dfrac{3}{10}$$

$$= \dfrac{6}{8}x + \dfrac{1}{3}y - \dfrac{3}{10}$$

Reduce the fractions, if possible:

$$= \frac{3}{4}x + \frac{1}{3}y - \frac{3}{10}$$

8   Answer: $1.3c^3 - 4.3c^2 + 2c + 7$

$1.3c^3 + 4.8 - 4.3c^2 + 2c + 2.2$

$= (1.3)c^3 + (-4.3)c^2 + (2)c + (4.8 + 2.2)$

$= 1.3c^3 - 4.3c^2 + 2c + 7$

Multiplying by Monomials

When multiplying terms, you will utilize multiplication and properties of exponents for like bases. For example, when multiplying $(3x^4)(4x^2)$, which have the same base of x, first multiply the coefficients, 3*4, which is 12. Then multiply $x^4*x^2 = x^6$, according to the Product Rule, for an overall answer of $12x^6$.

If multiplying a monomial and a polynomial, multiply the monomial by each of the terms in the polynomial.
For example, $2x^2 (4x^3 + 5x^2 - 3x)$. The first step is to multiply (*or distribute*) the outside term by each term of the polynomial $(2x^2)(4x^3) + (2x^2)(5x^2) - (2x^2)(3x)$. Following the same steps as the above example, the result is $8x^5 + 10x^4 - 6x^3$.

Here are 8 Practice problems for you:

Evaluate each expression.

1 $(a^{13}b^4)(12ab^6)$

2 $(-5u^2v)(-8u^3v^2)$

3 $2t(4t - 3)$

4 $-3a^2(-4a^2 + 2a - \dfrac{1}{3})$

5 $8pq(2pq - 3p + 5q)$

6 $5a^2b^3(2ab + 6a^3 - 3b^2)$

7 $-4u^2v(2u - 5uv^3 + v)$

8 $(k^2 - 13k - 6)(-4k)$

Answers and explanations:

1 Answer: $12a^{14}b^{10}$

$(a^{13}b^4)(12ab^6)$
Put like bases together:
$= (1*12)*(a^{13}*a)(b^4*b^6)$ *Remember a*
has an invisible exponent of 1
$= 12a^{(13+1)}b^{(4+6)}$
$= 12a^{14}b^{10}$

2 Answer: $40u^5v^3$

$(-5u^2v)(-8u^3v^2)$
$= [-5*(-8)](u^2*u^3)(v*v^2)$

$$= 40u^5v^3$$

3   Answer: $8t^2 - 6t$

$$2t(4t - 3)$$
$$= 2t*(4t) - 2t*(3)$$
$$= (2*4)(t*t) - 2*3(t)$$
$$= 8t^2 - 6t$$

4   Answer: $12a^4 - 6a^3 + a^2$

$$-3a^2(-4a^2 + 2a - \frac{1}{3})$$

$$= (-3a^2)(-4a^2) + (-3a^2)(2a) - (-3a^2)(\frac{1}{3})$$
$$= [-3*(-4)](a^2*a^2) + (-3*2)(a^2*a) -$$
$$(-3*\frac{1}{3})(a^2)$$
$$= 12a^4 - 6a^3 + a^2$$

5   Answer: $16p^2q^2 - 24p^2q + 40pq^2$

$$8pq (2pq - 3p + 5q)$$
$$= (8pq)(2pq) - (8pq)(3p) + (8pq)(5q)$$
$$= (8*2)(p*p)(q*q) - (8*3)(p*p)(q) +$$
$$(8*5)(p)(q*q)$$
$$= 16p^2q^2 - 24p^2q + 40pq^2$$

6   Answer: $10a^3b^4 + 30a^5b^3 - 15a^2b^5$

$5a^2b^3 (2ab + 6a^3 - 3b^2)$

$= (5a^2b^3)(2ab) + (5a^2b^3)(6a^3) -$
$\quad (5a^2b^3)(3b^2)$

$= (5*2)(a^2*a)(b^3*b) +$
$\quad (5*6)(a^2*a^3)(b^3) -$
$\quad (5*3)(a^2)(b^3*b^2)$

$= 10a^3b^4 + 30a^5b^3 - 15a^2b^5$

7   Answer: $-8u^3v + 20u^3v^4 - 4u^2v^2$

$-4u^2v (2u - 5uv^3 +v)$

$= (-4u^2v)(2u) - (-4u^2v)(5uv^3) +$
$\quad (-4u^2v)(v)$

$= (-4*2)(u^2*u)(v) - (-4*5)(u^2*u)$
$\quad (v*v^3) + (-4)(u^2)(v*v)$

$= -8u^3v + 20u^3v^4 - 4u^2v^2$

8   Answer: $-4k^3 + 52k^2 + 24k$

$(k^2 - 13k - 6)(-4k)$

Even though the monomial is in the back, you distribute it the same way as one that is listed first.

$= (k^2)(-4k) - (13k)(-4k) - (6)(-4k)$

$= - 4k^3 + 52k^2 + 24k$

Multiplying by Polynomials

We just finished multiplying a polynomial with a monomial. Multiplying a polynomial by a polynomial is an expansion of that and can take on several forms. You can multiply a binomial by a binomial, a binomial by a polynomial, and a polynomial by a polynomial.

When multiplying a binomial by a binomial, one method that can be employed is the FOIL method.

FOIL is an acronym for First, Outer, Inner, Last. Looking at the binomials $(c + 6)(c - 2)$, First means multiply the first term of each binomial, $c*c = c^2$. Outer means multiply the outer terms of the binomials, $c*(-2) = -2c$. For Inner, you multiply the inner most terms of the binomials, $6*c = 6c$. Last is for multiplying the latter terms of the binomials, $6*(-2) = -12$. After the products are determined, list them in polynomial form, combining any like terms, $-2c$ and $6c$, totaling $4c$. The resulting polynomial for this example is $c^2 + 4c - 12$. In most cases, you can expect the outer and inner products to combine. As a side note, the FOIL method can _only_ be used when multiplying a binomial by a binomial. For other types, you can utilize the methods shown next.

The Distributive method. This method utilizes the distributive property in order to find the products, which is simply multiplication. Let's use another example to display this method: $(x + 3)(x + 8)$. First, multiply the first term of the first binomial with the entire second binomial: $x(x + 8) = x*x + x*8 = x^2 + 8x$. Then multiply the second term of the first binomial with the entire second binomial: $3(x + 8) = 3*x + 3*8 =$

3x + 24. Lastly, combine the like terms of the products:
$x^2 + 8x + 3x + 24 = x^2 + 11x + 24$.

 Table Method. One additional method for multiplying polynomials is the Table method. Using another example, $(y - 5)(y - 2)$, the terms are input into the table and multiplied together to get the products.

*	y	–5
y	$y*y = y^2$	$-5*y = -5y$
–2	$-2*y = -2y$	$-5*(-2) = 10$

Once you've found all of the products, combine the like terms, –5y and –2y. The combined polynomial is $y^2 - 5y - 2y + 10$ equaling $y^2 - 7y + 10$.

When multiplying binomials by binomials, you can utilize any of the above methods. But when multiplying binomials by polynomials or polynomials by polynomials, you can only use the **Distributive method** or the **Table method.**

Binomial Practice Problems:

Use the method of your choice (FOIL, Distributive, or Table) to evaluate the expressions:

1   $(y - 10)(y + 9)$

2   $(x + 4)(x - 6)$

3   $(m - 12)(m - 2)$

4   $(n - 7)(n - 2)$

5   $(3p - 2)(4p + 1)$

6   $(7q + 11)(q - 5)$

7   $(-4w + 8)(-3w + 2)$

8   $(p - 3w)(p - 11w)$

Answers and explanations:

1   Answer: $y^2 - y - 90$

$(y - 10)(y + 9)$

Using FOIL:

$$F = y * y = y^2$$
$$O = 9 * y = 9y$$
$$I = -10 * y = -10y$$
$$L = -10 * 9 = -90$$

Combining results: $y^2 + 9y - 10y - 90$
$= y^2 - y - 90$

2   Answer: $x^2 - 2x - 24$

$(x + 4)(x - 6)$
Using FOIL:
$F = x * x = x^2$
$O = -6 * x = -6x$
$I = 4 * x = 4x$
$L = -4 * 6 = -24$

Combining results: $x^2 - 6x + 4x - 24$
$= x^2 - 2x - 24$

3   Answer: $m^2 - 14m + 24$

    $(m - 12)(m - 2)$
    Using FOIL:
        $F = m * m = m^2$
        $O = -2 * m = -2m$
        $I = -12 * m = -12m$
        $L = -12 * (-2) = 24$

    Combined: $m^2 - 2m - 12m + 24$
             $= m^2 - 14m + 24$

4   Answer: $n^2 - 9n + 14$

    $(n - 7)(n - 2)$
    Using Distributive Method:
        $n(n - 2) = (n * n) - (2 * n) = n^2 - 2n$
      $-7(n - 2) = (-7 * n) - [(-7) * (2)]$
               $= -7n - (-14)$
               $= -7n + 14$

    Combined: $n^2 - 2n - 7n + 14$
           $= n^2 - 9n + 14$

5   Answer: $12p^2 - 5p - 2$

    $(3p - 2)(4p + 1)$
    Using Distributive Method:
    $3p(4p + 1) = (3p * 4p) + (3p * 1) = 12p^2 + 3p$
    $-2(4p + 1) = (-2 * 4p) + ((-2) * 1) = -8p - 2$

    Combined: $12p^2 + 3p - 8p - 2$
           $= 12p^2 - 5p - 2$

6 Answer: $7q^2 - 24q - 55$

$(7q + 11)(q - 5)$
Using Distributive Method:

$7q\,(q - 5) = (7q * q) - (7q * 5) = 7q^2 - 35q$
$11\,(q - 5) = (11 * q) - (11 * 5) = 11q - 55$

Combined: $7q^2 - 35q + 11q - 55$
$\qquad\quad = 7q^2 - 24q - 55$

7 Answer: $12w^2 - 32w + 16$

$(-4w + 8)(-3w + 2)$
Using the Table Method:

*	−4w	8
−3w	−4w * (−3w) = $12w^2$	8 * (−3w) = −24w
2	−4w * 2 = −8w	8 * 2 = 16

$= 12w^2 - 24w - 8w + 16$
$= 12w^2 - 32w + 16$

8 Answer: $p^2 - 14pw + 33w^2$

$(p - 3w)(p - 11w)$
Using Table Method:

*	p	–3w
p	p * p = p²	–3w * p = –3w * p = –3pw
–11w	p * (–11w) = –11pw	–3w * (–11w) = 33w²

$$= p^2 - 3pw - 11pw + 33w^2$$
$$= p^2 - 14pw + 33w^2$$

Let's do some more examples.

Multiplying a Binomial by a polynomial
$$(3x + 4)(2x^3 - 5x^2 + 6x)$$

Using the Distributive method:
$3x (2x^3 - 5x^2 + 6x)$
$= 3x*2x^3 + 3x*(-5x^2) + 3x*6x$
$= 6x^4 - 15x^3 + 18x^2$

Then, $4(2x^3 - 5x^2 + 6x)$
$= 4*2x^3 + 4(-5x^2) + 4*6x$
$= 8x^3 - 20x^2 + 24x$

Lastly combine the results and like terms:
$6x^4 - 15x^3 + 18x^2 + 8x^3 - 20x^2 + 24x$
$= 6x^4 - 7x^3 - 2x^2 + 24x$

Here is another example. This time we're using the **Table method**:

$$(y - 2)(3y^3 + y - 5)$$

26

*	y	–2
3y³	$y * 3y^3 = 3y^4$	$-2 * 3y^3 = -6y^3$
y	$y * y = y^2$	$-2 * y = -2y$
–5	$y * (-5) = -5y$	$-5 * (-2) = 10$

Now putting together the string of the polynomial, $3y^4 - 6y^3 + y^2 - 2y - 5y + 10$

Combine like terms to yield: $3y^4 - 6y^3 + y^2 - 7y + 10$

These methods can also be utilized when multiplying a polynomial by a polynomial. We will do an example with the Table method.

$(2x^2 - 3x - 4)(3x^2 - 5x + 1)$

*	$3x^2$	$-5x$	1
$2x^2$	$2x^2 * 3x^2$ $= 6x^4$	$2x^2 * (-5x)$ $= -10x^3$	$2x^2 * 1$ $= 2x^2$
$-3x$	$-3x * 3x^2$ $= -9x^3$	$-3x * (-5x)$ $= 15x^2$	$-3x * 1$ $= -3x$
-4	$-4 * 3x^2$ $= -12x^2$	$-4 * (-5x)$ $= 20x$	$-4 * 1$ $= -4$

Putting all of the products together:

$$6x^4 - 10x^3 + 2x^2 - 9x^3 + 15x^2 - 3x - 12x^2 + 20x - 4$$

Combining like terms:

$$= 6x^4 - 10x^3 - 9x^3 + 2x^2 + 15x^2 - 12x^2 - 3x + 20x - 4$$
$$= 6x^4 - 19x^3 + 5x^2 + 17x - 4$$

Here are a few practice problems to do.

Simplify.

1) $(5s + 3)(s^2 + s - 2)$
2) $(t - 4)(2t^2 - t + 6)$
3) $(2a - 5)(3a^2 - 4a + 9)$
4) $(3w - 2)(9w^2 + 6w + 4)$

Answers and explanation:

1) Answer: $5s^3 + 8s^2 - 7s - 6$

   By distributing:
   $$5s(s^2 + s - 2) + 3(s^2 + s - 2)$$
   $$= (5s * s^2) + (5s * s) - (5s * 2) + (3 * s^2)$$
   $$+ (3 * s) - (3 * 2)$$
   $$= 5s^3 + 5s^2 - 10s + 3s^2 + 3s - 6$$
   $$= 5s^3 + 8s^2 - 7s - 6$$

2) Answer: $2t^3 - 9t^2 + 10t - 24$

By distributing:
$t(2t^2 - t + 6) - 4(2t^2 - t + 6)$
$= (t * 2t^2) - (t * t) + (t * 6) - (4 * 2t^2) - (4 * (-t))$
$\quad - (4 * 6)$
$= 2t^3 - t^2 + 6t - 8t^2 + 4t - 24$
$= 2t^3 - 9t^2 + 10t - 24$

3) Answer: $6a^3 - 23a^2 + 38a - 45$

By table method:

*	2a	–5
$3a^2$	$2a * 3a^2 = 6a^3$	$-5 * 3a^2 = -15a^2$
$-4a$	$2a * (-4a) = -8a^2$	$-5 * (-4a) = 20a$
9	$2a * 9 = 18a$	$-5 * 9 = -45$

$= 6a^3 - 15a^2 - 8a^2 + 20a + 18a - 45$
$= 6a^3 - 23a^2 + 38a - 45$

4) Answer: $27w^3 - 8$

By table method:

*	3w	−2
$9w^2$	$3w * 9w^2 = 27w^3$	$-2 * 9w^2 = -18w^2$
6w	$3w * 6w = 18w^2$	$-2 * 6w = -12w$
4	$3w * 4 = 12w$	$4 * (-2) = -8$

$$= 27w^3 - 18w^2 + 18w^2 - 12w + 12w - 8$$
$$= 27w^3 - 8$$

Special Products

Special products includes the Difference of Squares, Perfect Square Trinomials, and Cubics.

Difference of Squares

Using any of the previously discussed methods, the expansion of $(2x + 3)(2x - 3)$ results in $4x^2 - 6x + 6x - 9$. After the like terms are combined, $4x^2 - 9$ is what remains.

The product results in the difference of the square of the first term and the square of the second term. You can either go through the steps of multiplying out the products or remember/memorize the difference of squares result.

Perfect Square Trinomials

Perfect square trinomials is the result when a binomial is squared, like $(3x + 5)^2$ or $(x - 6)^2$.

Finding the product of $(3x + 5)^2$ is $(3x + 5)(3x + 5)$ which, when expanded, yields $9x^2 + 15x + 15x + 25$. After the like terms are combined, you have $9x^2 + 30x + 25$. This is the sum of the square of the first term, twice the product of the first and second terms, and the square of the second term. So $(a + b)^2 = a^2 + 2ab + b^2$.

A very common error happens here. Many times people just square the first term and then the second term and add them together. DO NOT make this mistake. Because the binomial is being squared, there will be a middle term.

By the same token, the product of binomials $(x - 6)^2$ is $x^2 - 12x + 36$. This is the first term squared minus twice the product of the first and second terms plus the square of the second term, $(a - b)^2 = a^2 - 2ab + b^2$. Notice the difference in the perfect square formulas. If the binomial includes addition, then the middle term will be positive. Also, if the binomial includes subtraction, then the middle term will be negative.

Examples to work:

1 $(3a - 4b)(3a + 4b)$

2 $(5y + 7x)(5y - 7x)$

3 $(2h - 5)(2h + 5)$

4     $(\frac{1}{2} - t)(\frac{1}{2} + t)$

5   $(a + 5)^2$

6   $(5d - 9)^2$

7   $(u^2 + 4v)^2$

8   $(3t^2 - 4s)^2$

Answers and explanations:

1   Answer: $9a^2 - 16b^2$

$(3a - 4b)(3a + 4b)$

$= 9a^2 + 12ab - 12ab - 16b^2$

$= 9a^2 - 16b^2$

2   Answer: $25y^2 - 49x^2$

$(5y + 7x)(5y - 7x)$

$= 25y^2 - 35xy + 35xy - 49x^2$

$= 25y^2 - 49x^2$

3   Answer: $4h^2 - 25$

$(2h - 5)(2h + 5)$

$= 4h^2 + 10h - 10h - 25$

$= 4h^2 - 25$

4 Answer: $\dfrac{1}{4} - t^2$

$$(\dfrac{1}{2} - t)(\dfrac{1}{2} + t)$$

$$= \dfrac{1}{4} + \dfrac{1}{2}t - \dfrac{1}{2}t - t^2$$

$$= \dfrac{1}{4} - t^2$$

5 Answer: $a^2 + 10a + 25$

$(a+5)^2$

$= a^2 + 5a + 5a + 25$

$= a^2 + 10a + 25$

6 Answer: $25d^2 - 90d + 81$

$(5d - 9)^2$

$= 25d^2 - 45d - 45d + 81$

$= 25d^2 - 90d + 81$

7 Answer: $u^4 + 8u^2v + 16v^2$

$(u^2 + 4v)^2$

$= u^4 + 4u^2v + 4u^2v + 16v^2$

$= u^4 + 8u^2v + 16v^2$

8   Answer: $9t^4 - 24t^2s + 16s^2$

$$(3t^2 - 4s)^2$$
$$= 9t^4 - 12t^2s - 12t^2s + 16s^2$$
$$= 9t^4 - 24t^2s + 16s^2$$

Cubics

When finding the cube of a binomial, there is a formula similar to the perfect square trinomials.

$$(x + y)^3 = x^3 + 3x^2y + 3xy^2 + y^3$$

and

$$(x - y)^3 = x^3 - 3x^2y + 3xy^2 - y^3.$$

You can memorize and use the above formulas or, alternatively, you can work it out by squaring the binomial first then multiply the resulting trinomial with the original binomial using the Table or Distributive method. After combining any like terms you'll have the answer.

Let's work an example, using the formulas.

$$(2x + 5y)^3$$

First term: 2x
Second term: 5y

First term cubed: $(2x)^3 = (2)^3 * x^3 = 8x^3$

3*first term squared*second term
$= 3*(2x)^2*(5y) = 3 * 4x^2 * 5y = (3*4*5)(x^2y)$
$= 60x^2y$

3*first term*second term squared
$= 3*(2x)*(5y)^2 = 3 * 2x * 25y^2 = (3*2*25)xy^2$
$= 150xy^2$

Second term cubed: $(5y)^3 = 5^3 * y^3 = 125y^3$

Combining all resulting terms:
$$8x^3 + 60x^2y + 150xy^2 + 125y^3$$

Cubics practice problems:

1 $(x + 2)^3$

2 $(3x - 2)^3$

3 $(5x + 4y)^3$

Answers and explanations
1 Answer: $x^3 + 6x^2 + 12x + 8$

$x^3 + (3 * x^2 * 2) + (3 * x * 2^2) + 2^3$
$= x^3 + 6x^2 + (3 * x * 4) + 8$
$= x^3 + 6x^2 + 12x + 8$

2 Answer: $27x^3 - 54x^2 + 36x - 8$

$(3x)^3 - [3 * (3x)^2 * 2] + (3 * 3x * 2^2) - 2^3$

$$=)3^3 * x^3) - (3 * 3^2 * x^2 * 2) + (3 * 3x * 4) - 8$$
$$= 27x^3 - (6 * 9 * x^2) + 36x - 8$$
$$= 27x^3 - 54x^2 + 36x - 8$$

3   Answer: $125x^3 + 300x^2y + 240xy2 + 64y3$

$(5x)^3 + [3 * (5x)^2 * 4y] + [3 * 5x * (4y)^2] + (4y)^3$

$$= (5^3 * x^3) + (3 * 5^2 * x^2 * 4y) + (3 * 5x * 4^2 * y^2) + (4^3 * y^3)$$

$$= 125x^3 + (3*25*4*x^2y) + (3*5*16xy^2) + 64y^3$$

$$= 125x^3 + 300x^2y + 240xy^2 + 64y^3$$

Dividing Polynomials

Division of polynomials includes dividing a polynomial by a monomial and dividing a polynomial by a binomial. The latter involves using long division.

Dividing by a Monomial

When dividing a polynomial by a monomial, you divide each individual term of the polynomial by the monomial in the denominator. Then you simplify the resulting terms.

Let's work through an example:

$$\frac{5a^3 - 10a^2 + 20a}{5a}$$

Divide each term by the denominator individually.

$$\frac{5a^3}{5a} - \frac{10a^2}{5a} + \frac{20a}{5a}$$

Now reduce each term to simplest form.

$$= a^2 - 2a + 4$$

Let's work some practice problems:

1   $(6x^2 + 4x - 14) \div (2)$

2   $(36a^4 - 48a^3 + 12a^2) \div (6a^3)$

3   $(3p^3 - p^2) \div (p)$

4   $(12y^2z^3 - 15yz^2 + 6y^2z) \div (-6y^2z)$

5   $(-15x^3y^4 + 25x^2y^3 - 5xy^2) \div (-5xy^2)$

6   $(4m^2 + 8m) \div (4m^2)$

7   $\dfrac{20w^3 + 15w^2 - w + 5}{10w}$

8   $\dfrac{2y^3 - 2y^2 + 3y - 9}{2y^2}$

Answers and explanations:

1   Answer: $3x^2 + 2x - 7$

$(6x^2 + 4x - 14) \div (2)$

Divide each term by 2 and reduce

$$\frac{6x^2}{2} + \frac{4x}{2} - \frac{14}{2}$$

$$= 3x^2 + 2x - 7$$

2

Answer: $6a - 8 + \dfrac{2}{a}$

$$(36a^4 - 48a^3 + 12a^2) \div (6a^3)$$

$$= \frac{36a^4}{6a^3} - \frac{48a^3}{6a^3} + \frac{12a^2}{6a^3}$$

$$= 6a - 8 + \frac{2}{a}$$

*Notice how the third term $\dfrac{a^2}{a^3}$ reduces leaving a in the denominator of the 3rd term.

3   Answer: $3p^2 - p$

$$(3p^3 - p^2) \div (p)$$

$$= \frac{3p^3}{p} - \frac{p^2}{p}$$

$$= 3p^2 - p$$

4
$$Answer: \; -2z^2 + \frac{5z}{2y} - 1$$

$$(12y^2z^3 - 15yz^2 + 6y^2z) \div (-6y^2z)$$

$$= \frac{12y^2z^3}{-6y^2z} - \frac{15yz^2}{-6y^2z} + \frac{6y^2z}{-6y^2z}$$

$$= -2z^2 + \frac{5z}{2y} - 1$$

5   Answer: $3x^2y^2 - 5xy + 1$

$$(-15x^3y^4 + 25x^2y^3 - 5xy^2) \div (-5xy^2)$$

$$= \frac{-15x^3y^4}{-5xy^2} + \frac{25x^2y^3}{-5xy^2} - \frac{5xy^2}{-5xy^2}$$

$$= 3x^2y^2 - 5xy + 1$$

6
$$Answer: \; 1 + \frac{2}{m}$$

$$(4m^2 + 8m) \div (4m^2)$$

$$= \frac{4m^2}{4m^2} + \frac{8m}{4m^2}$$

$$= 1 + \frac{2}{m}$$

7

Answer: $2w^2 + \dfrac{3}{2}w - \dfrac{1}{10} + \dfrac{1}{2w}$

$$\dfrac{20w^3 + 15w^2 - w + 5}{10w}$$

$$= \dfrac{20w^3}{10w} + \dfrac{15w^2}{10w} - \dfrac{w}{10w} + \dfrac{5}{10w}$$

$$= 2w^2 + \dfrac{3}{2}w - \dfrac{1}{10} + \dfrac{1}{2w}$$

Notice how 1 remains as the numerator of the 3rd term once the w's are reduced and in the 4th term when 5/10 is reduced.

8

Answer: $y - 1 + \dfrac{3}{2y} - \dfrac{9}{2y^2}$

$$\dfrac{2y^3 - 2y^2 + 3y - 9}{2y^2}$$

$$= \dfrac{2y^3}{2y^2} - \dfrac{2y^2}{2y^2} + \dfrac{3y}{2y^2} - \dfrac{9}{2y^2}$$

$$= y - 1 + \frac{3}{2y} - \frac{9}{2y^2}$$

Dividing Polynomials By Polynomials

To divide polynomials by polynomials, we use long division.

Let's step through the process with an example. Note: this is the same process as regular division, without variables.

$(2x^2 - x + 3) \div (x - 3)$

The first step is to divide the first term of the polynomial, $2x^2$, by the first term of the monomial, x. This yields 2x.

After you've divided the first term and found the quotient, 2x, then multiply that by both terms of the binomial. Notice that you should place the terms under their like terms in order to subtract them.

$$
\begin{array}{r}
2x \\
x - 3 \overline{)\, 2x^2 - x\ +\ 3} \\
2x^2 - 6x
\end{array}
$$
   *This is to be subtracted from the polynomial. One way to do this is to change the signs of the resulting $2x^2 - 6x$ to $-2x^2 + 6x$.

Result:

$$
\begin{array}{r}
2x \\
x - 3 \overline{)\, 2x^2\ -\ x\ +\ 3}
\end{array}
$$

41

$$\underline{-2x^2 + 6x}$$
$$0 \quad + 5x$$

* Bring down the next term and repeat

$$x - 3 \overline{)\ 5x + 3}$$

Next, find quotient of 5x divided by x. It is 5.

$$\begin{array}{r} 5 \\ x - 3 \overline{)\ 5x + 3} \\ \underline{5x - 15} \end{array}$$

*The result of multiplying 5 times the binomial.

$$\begin{array}{r} 5 \\ x - 3 \overline{)\ 5x + 3} \\ \underline{-5x + 15} \\ 0 \ + 18 \end{array}$$

*Change the sign and add

Once you've divided all of the terms, the remainder is put over the original divisor binomial. Our remainder in this example is 18.

So the answer is $2x + 5 + \dfrac{18}{x - 3}$

Here's one more example before you do some practice problems.

$(3x^2 + 2x - 5) \div (x + 2)$

$$\begin{array}{r} 3x - 4 \\ x + 2 \overline{)\ 3x^2 + 2x - 5} \\ \underline{-3x^2 - 6x} \end{array}$$

* Change signs, add, and bring down the next term

42

$$-4x - 5 \qquad \text{* Divide x into } -4x$$
$$\underline{4x + 8} \qquad \text{* Change signs and add}$$
$$3 \qquad \text{* This is the remainder}$$

Final answer is:

$$3x - 4 + \dfrac{3}{x + 2}$$

Sometimes you'll have polynomials with missing terms. In this case, you must first fill them in with 0's as their coefficients. For example, when doing $(2w^3 + 8w^2 - 16) \div (2w + 4)$, you have to change the first polynomial to $(2w^3 + 8w^2 + 0w - 16) \div (2w + 4)$.

$$
\begin{array}{r}
w^2 + 2w - 4 \\
\hline
2w + 4 \,)\, 2w^3 + 8w^2 + 0w - 16 \\
\underline{-2w^3 - 4w^2}
\end{array}
$$
* $2w^3$ divided by
  $2w$ is w^2
* Change signs, add, and bring down next term

$$4w^2 + 0w \qquad \text{* } 4w^2 \div 2w \text{ is } 2w$$
$$\underline{- 4w^2 - 8w} \qquad \text{* Change signs and add}$$

$$- 8w - 16 \qquad \text{* Divide by } 2w$$
$$\underline{8w + 16}$$
$$0 \qquad \text{* No remainder}$$

Final answer: $w^2 + 2w - 4$

Practice problems:

1. $\dfrac{t^2 + 4t + 5}{t + 1}$

2. $\dfrac{7b^2 - 3b - 4}{b - 1}$

3    $\dfrac{5k^2 - 29k - 6}{5k + 1}$

4    $\dfrac{4p^3 + 12p^2 + p - 12}{2p + 3}$

5    $\dfrac{12a^3 - 2a^2 - 17a - 5}{3a + 1}$

6    $\dfrac{4x^3 - 3x - 26}{x - 2}$

7    $\dfrac{9x^3 + 11x + 10}{3x + 2}$

8    $\dfrac{w^4 + 5w^3 - 5w^2 - 15w + 7}{w^2 - 3}$

Answers and explanations:

1   *Answer:* $t + 3 + \dfrac{2}{t + 1}$

$$
\begin{array}{r}
t + 3 \\
t + 1 \,\overline{)\, t^2 + 4t + 5} \\
\underline{t^2 + t} \\
3t + 5 \\
\underline{3t + 3} \\
2
\end{array}
$$

Answer: $t + 3 + \dfrac{2}{t + 1}$

2   Answer: $7b + 4$

$$
\begin{array}{r}
7b + 4 \\
b - 1 \,\big)\, \overline{7b^2 - 3b - 4} \\
\underline{7b2 - 7b} \\
4b - 4 \\
\underline{4b - 4} \\
0
\end{array}
$$

3   Answer: $k - 6$

$$
\begin{array}{r}
k - 6 \\
5k + 1 \,\big)\, \overline{5k^2 - 29k - 6} \\
\underline{5k^2 + \ \ k} \\
-30k - 6 \\
\underline{-30k - 6} \\
0
\end{array}
$$

4   Answer: $2p^2 + 3p - 4$

$$
\begin{array}{r}
2p2 + 3p - 4 \\
2p + 3 \,\big)\, \overline{4p^3 + 12p^2 + p - 12} \\
\underline{4p^3 + \ 6p^2} \\
6p^2 + \ p \\
\underline{6p^2 + 9p} \\
-8p - 12 \\
\underline{-8p - 12} \\
0
\end{array}
$$

5   Answer: $4a^2 - 2a - 5$

$$
\begin{array}{r}
4a^2 - 2a - 5 \\
3a + 1 \,)\, \overline{12a^3 - 2a^2 - 17a - 5} \\
\underline{12a^3 + 4a^2} \\
-6a^2 - 17a \\
\underline{-6a^2 - 2a} \\
-15a - 5 \\
\underline{-15a - 5} \\
0
\end{array}
$$

6   Answer: $4x^2 + 8x + 13$

$$
\begin{array}{r}
4x^2 + 8x + 13 \\
x - 2 \,)\, \overline{4x^3 + 0x^2 - 3x - 26} \\
\underline{4x^3 - 8x^2} \\
8x^2 - 3x \\
\underline{8x^2 - 16x} \\
13x - 26 \\
\underline{13x - 26} \\
0
\end{array}
$$

7   Answer: $3x^2 - 2x + 5$

$$
\begin{array}{r}
3x^2 - 2x + 5 \\
3x + 2 \,)\, \overline{9x^3 + 0x^2 + 11x + 10} \\
\underline{9x^3 + 6x^2} \\
-6x^2 + 11x \\
\underline{-6x^2 - 4x} \\
15x + 10 \\
\underline{15x + 10} \\
0
\end{array}
$$

8   Answer: $w^2 + 5w - 2 + \dfrac{1}{w^2 - 3}$

$$w^2 + 5w - 2 + \dfrac{1}{w^2 - 3}$$

$$w^2 - 3 \overline{) \; w^4 + 5w^3 - 5w^2 - 15w + 7}$$

$$\underline{w^4 \qquad\quad - 3w^2}$$

$$5w3 - 2w^2 - 15w$$

$$\underline{5w^3 \qquad\quad - 15w}$$

$$-2w^2 \qquad\quad + 7$$

$$\underline{-2w^2 \qquad\quad + 6}$$

$$1$$

Applications using Polynomials

Polynomials can be employed in their use in applications to geometry. One of these includes perimeter.

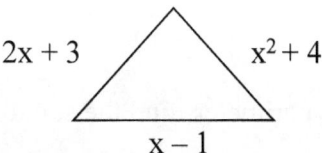

2x + 3 $x^2 + 4$

x − 1

Remember, to find the perimeter, you add all the sides of a shape.

Perimeter = side 1 + side 2 + side 3

$$P = (2x + 3) + (x^2 + 4) + (x - 1)$$

*Order the terms from highest exponent to least

$$P = x^2 + 2x + x + 3 + 4 - 1$$

$$P = x^2 + 3x + 6$$

Another example of this use:

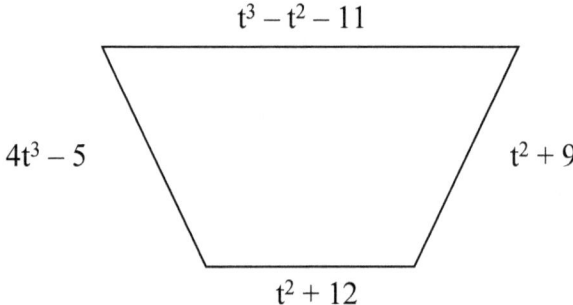

To obtain the perimeter, find the sum of the sides.

$$P = (t^3 - t^2 - 11) + (4t^3 - 5) + (2t^2 + 9) + (t + 12)$$

$$P = t^3 + 4t^3 - t^2 + 2t^2 + t - 11 - 5 + 9 + 12$$

$$P = 5t^3 + t^2 + t + 5$$

Polynomials can be used in finding area, also.

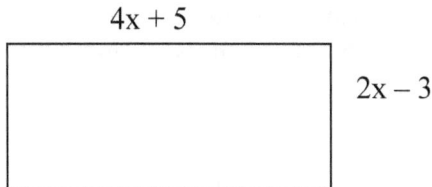

4x + 5

2x − 3

Area of a rectangle is found by multiplying the length by the width.

Area = (4x + 5) * (2x − 3)

$A = 8x^2 - 2x - 15$

In Closing

This book has covered every aspect of working with polynomials. I do hope you found it comprehensive and easy to follow. It may have taken you longer than 8 minutes to master all the skills, but if you've followed the steps and practiced the problems provided, you should be able to work with polynomials in any type of problem with confidence.

Other Titles by Arlissa Pinkelton:

Learn To Factor in 8 Minutes
Learn About Exponents in 8 Minutes

(Coming soon)
Learn Radicals in 8 Minutes

Learn Rational Expressions in 8 Minutes
Learn About Inequalities in 8 Minutes
Solving & Graphing Linear Equations in 8 Minutes
Learn All You Need To Know About Real Numbers in 8 Minutes

www.ingramcontent.com/pod-product-compliance
Lightning Source LLC
Chambersburg PA
CBHW071006180526
45168CB00003B/1314